United States Government Accountabili[...]

Report to Congressional [...]

I0011642

April 2016

CLOUD COMPUTING

Agencies Need to Incorporate Key Practices to Ensure Effective Performance

GAO Highlights

Highlights of GAO-16-325, a report to congressional requesters

April 2016

CLOUD COMPUTING

Agencies Need to Incorporate Key Practices to Ensure Effective Performance

Why GAO Did This Study

Cloud computing is a means for delivering computing services via IT networks. When executed effectively, cloud-based services can allow agencies to pay for only the IT services used, thus paying less for more services. An important element of acquiring cloud services is a service level agreement that specifies, among other things, what services a cloud provider is to perform and at what level.

GAO was asked to examine federal agencies' use of SLAs. GAO's objectives were to (1) identify key practices in cloud computing SLAs and (2) determine the extent to which federal agencies have incorporated such practices into their SLAs. GAO analyzed research, studies, and guidance developed by federal and private entities to develop a list of key practices to be included in SLAs. GAO validated its list with the entities, including OMB, and analyzed 21 cloud service contracts and related documentation of five agencies (with the largest fiscal year 2015 IT budgets) against the key practices to identify any variances, their causes, and impacts.

What GAO Recommends

GAO recommends that OMB include all ten key practices in future guidance to agencies and that Defense, Health and Human Services, Homeland Security, Treasury, and Veterans Affairs implement SLA guidance and incorporate applicable key practices into their SLAs. In commenting on a draft of this report, OMB and one agency had no comment, the remaining four agencies concurred with GAO's recommendations.

View GAO-16-325. For more information, contact David A. Powner at (202) 512-9286 or pownerd@gao.gov.

What GAO Found

Federal and private sector guidance highlights the importance of federal agencies using a service level agreement (SLA) in a contract when acquiring information technology (IT) services through a cloud computing services provider. An SLA defines the level of service and performance expected from a provider, how that performance will be measured, and what enforcement mechanisms will be used to ensure the specified performance levels are achieved. GAO identified ten key practices to be included in an SLA, such as identifying the roles and responsibilities of major stakeholders, defining performance objectives, and specifying security metrics. The key practices, if properly implemented, can help agencies ensure services are performed effectively, efficiently, and securely. Under the direction of the Office of Management and Budget (OMB), guidance issued to agencies in February 2012 included seven of the ten key practices described in this report that could help agencies ensure the effectiveness of their cloud services contracts.

GAO determined that the five agencies and the 21 cloud service contracts it reviewed had included a majority of the ten key practices. Specifically, of the 21 cloud service contracts reviewed from the Departments of Defense, Health and Human Services, Homeland Security, Treasury, and Veterans Affairs, 7 had fulfilled all 10 of the key practices, as illustrated in the figure. The remaining 13 contracts had incorporated 5 or more of the 10 key practices and 1 had not included any practices.

Figure 1: Number of Cloud Service Contracts That Met All 10 Key Practices

Agencies

Total cloud service contracts reviewed ▮ Number of contracts that met all key practices

Source: GAO analysis of agency data. | GAO-16-325

Agency officials gave several reasons for why they did not include all elements of the key practices into their cloud service contracts, including that guidance directing the use of such practices had not been created when the cloud services were acquired. Unless agencies fully implement SLA key practices into their SLAs, they may not be able to adequately measure the performance of the services, and, therefore, may not be able to effectively hold the contractors accountable when performance falls short.

_____ United States Government Accountability Office

Contents

Figures

Abbreviations

CIO	chief information officer
DOD	Department of Defense
DHS	Department of Homeland Security
FAR	Federal Acquisition Regulation
GSA	General Services Administration
HHS	Department of Health and Human Services
IT	information technology
IRS	Internal Revenue Service
NIST	National Institute of Standards and Technology
OMB	Office of Management and Budget
SLA	service level agreement
Treasury	Department of the Treasury
VA	Department of Veterans Affairs

GAO

U.S. GOVERNMENT ACCOUNTABILITY OFFICE

441 G St. N.W.
Washington, DC 20548

April 7, 2016

The Honorable Ron Johnson
Chairman
The Honorable Thomas R. Carper
Ranking Member
Committee on Homeland Security and Governmental Affairs
United States Senate

The Honorable Jason Chaffetz
Chairman
Committee on Oversight and Government Reform
House of Representatives

Cloud computing is a process for acquiring and delivering computing services via internal or external information technology (IT) networks. According to the National Institute of Standards and Technology (NIST), cloud computing is "a means for enabling on-demand access to shared and scalable pools of computing resources with the goal of minimizing management effort or service provider interaction." More specifically, purchasing IT services through a provider enables agencies to avoid paying for all the assets (e.g., hardware, software, networks) that would typically be needed to provide such services. This approach offers federal agencies a means to buy the services faster and possibly cheaper than through the traditional methods they have used, such as keeping it all in-house. To take advantage of these potential benefits, agencies have reported that they plan to spend more than $2 billion on cloud computing services in fiscal year 2016.[1]

An important part of acquiring IT cloud computing services is incorporating a service level agreement (SLA) into the contract. An SLA defines levels of service and performance that the agency expects the contractor to meet and the agency uses the information to measure the effectiveness of its cloud services. To encourage the use of SLAs, the Office of Management and Budget (OMB) directed subject matter experts to issue guidance that highlighted SLAs as a key factor to be addressed

---

[1] https://www.itdashboard.gov.

in developing cloud computing contracts.[2] You asked us to examine federal agencies' use of SLAs. Specifically, our objectives were to (1) identify key practices used in cloud computing service level agreements to ensure service is performed at specified levels and (2) determine the extent to which federal agencies have incorporated such practices into their cloud computing service level agreements.

To identify key practices, we analyzed SLA research, studies, and guidance developed and used by federal agencies and private entities and performed a comparative analysis of the practices to identify the most effective ones. Specifically, we analyzed information from publications and related documentation issued by the following ten public and private organizations to determine key SLA practices:

- Federal Chief Information Officers Council
- Chief Acquisitions Officers Council
- National Institute of Standards and Technology
- European Commission Directorate General for Communications Networks, Content and Technology
- OMB
- Gartner
- MITRE Corporation
- Cloud Standards Customer Council
- International Organization for Standardization
- International Electrotechnical Commission

We then validated our analysis through interviews with experts from these organizations. We also had officials from OMB review and validate that the practices we identified are those the office expects federal agencies to follow.

To determine the extent to which federal agencies have incorporated key practices into their cloud computing contracts, we selected five agencies to review based, in part, on the size of their largest fiscal year 2015 IT budgets and planned spending on cloud computing services. The agencies selected were the Departments of

---

[2]Chief Information Officer's Council and Chief Acquisition Officers Council, *Creating Effective Cloud Computing Contracts for the Federal Government, Best Practices for Acquiring IT as a Service* (Feb. 24, 2012). The guidance was written in coordination with the Federal Cloud Compliance Committee.

GAO-16-325 Cloud Computing

- Defense,
- Health and Human Services,
- Homeland Security,
- Treasury, and
- Veterans Affairs.

To select and review the cloud services used by the agencies, we obtained a list of each agency's cloud services. We listed the cloud services for each agency and selected two for each of three major cloud service models (infrastructure, platform, or software).[3] In certain cases, the agency did not have two cloud services for a service model, so the number chosen for that service model was less than two.

For each of the selected cloud services, we compared its contract (if one existed), any supporting SLA, and other documentation to our list of key practices to determine the extent to which the agency had adhered to the key practices or if there were variances.[4] We also interviewed agency officials to corroborate our analysis and identify the causes and impacts of any variances from the practices. (Further details of our scope and methodology are in app. I.)

We conducted this performance audit from January 2015 to April 2016 in accordance to generally accepted government auditing standards. Those standards require that we plan and perform the audit to obtain sufficient, appropriate evidence to provide a reasonable basis for our findings and conclusions based on our audit objectives. We believe that the evidence obtained provides a reasonable basis for our findings and conclusions based on our audit objectives.

Background

The federal government spends more than $80 billion dollars on IT annually, with more than $2 billion of that amount spent on acquiring cloud-based services. This amount is expected to rise in coming fiscal years, according to OMB. A goal of these investments is to improve

---

[3]An expanded description of each of these types of service is included in the following section.

[4]Types of documentation included contract modifications, subscription agreement, terms of use, terms of condition, performance work statement, support agreement, service level compliance report, application management services, technical proposal, and customer agreements.

federal IT systems by replacing aging and duplicative infrastructure and systems that are costly and difficult to maintain. Cloud computing helps do this by giving agencies the ability to purchase a broad range of IT services in a utility-based model that allows an agency to pay for only the IT services it uses.[5]

According to NIST, an application should possess five essential characteristics to be considered cloud computing: on-demand self-service, broad network access, resource pooling, rapid elasticity, and measured service.[6] Essentially, cloud computing applications are network-based and scalable on demand.

According to OMB, cloud computing is economical, flexible, and fast:

- Economical: cloud computing can be a pay-as-you-go approach, in which a low initial investment is required to begin and additional investment is needed only as system use increases.

- Flexible: IT departments that anticipate fluctuations in user demand no longer need to scramble for hardware and software to meet increasing need. With cloud computing, capacity can be added or subtracted quickly.

- Fast: cloud computing eliminates long procurement and certification processes, while providing a wide selection of services.

In addition, according to NIST, cloud computing offers three service models:

- Infrastructure as a service—the agency has the capability to provision processing, storage, networks, and other fundamental computing

---

[5]Chief Information Officer's Council and the Chief Acquisition Officers Council, *Creating Effective Cloud Computing Contracts for the Federal Government* (Feb. 24, 2012).

[6]NIST defines these characteristics further. On-demand self-service allows consumers to acquire computing capabilities automatically and as needed. Broad network access provides capabilities over a network, which is accessed through standard mechanisms (e.g., a mobile phone, tablet, laptop, and workstation). Resource pooling means the vendor's combined computing resources serve multiple consumers. Rapid elasticity refers to the ability to vary resources commensurate with demand. Measured services are incrementally valued, typically on a pay-per-use, or charge-per-use, basis.

resources and run its own software, including operating systems and applications. The agency does not manage or control the underlying infrastructure but controls and configures operating systems, storage, deployed applications, and possibly, selected networking components (e.g., host firewalls).

- Platform as a service—the agency deploys its own or acquired applications created using programming languages and tools supported by the provider. The agency does not manage or control the underlying infrastructure, but controls and configures the deployed applications.

- Software as a service—the agency uses the service provider's applications, which are accessible from various client devices through an interface such as a Web browser (e.g., Web-based e-mail system). The agency does not manage or control the underlying infrastructure or the individual application capabilities.

As can be seen in figure 1, each service model offers unique functionality, with consumer control of the environment decreasing from infrastructure to platform to software.

Figure 1: Cloud Service Provider and Consumer Responsibilities for the Three Service Models

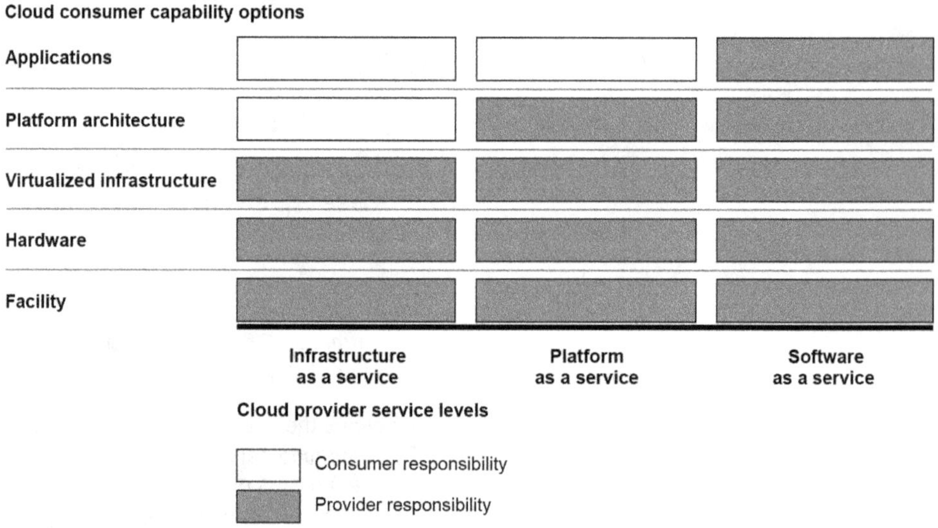

Cloud consumer capability options

	Infrastructure as a service	Platform as a service	Software as a service
Applications	Consumer	Consumer	Provider
Platform architecture	Consumer	Provider	Provider
Virtualized infrastructure	Provider	Provider	Provider
Hardware	Provider	Provider	Provider
Facility	Provider	Provider	Provider

Cloud provider service levels

☐ Consumer responsibility
▨ Provider responsibility

Source: GAO analysis based on National Institute of Standards and Technology information. | GAO-16-325

NIST has also defined four deployment models for providing cloud services: private, community, public, and hybrid.

- In a private cloud, the service is set up specifically for one organization, although there may be multiple customers within that organization and the cloud may exist on or off the customer's premises.

- In a community cloud, the service is shared by organizations with similar requirements. The cloud may be managed by the organizations or a third party and may exist on or off an organization's premises.

- A public cloud is available to the general public and is owned and operated by the service provider.

- A hybrid cloud is a composite of two or more other deployment models (private, community, or public) that are bound together by standardized or proprietary technology.

According to federal guidance, these deployment models determine the number of consumers and the nature of other consumers' data that may be present in a cloud environment. A public cloud should not allow a consumer to know or control other consumers of a cloud service provider's environment. However, a private cloud can allow for ultimate control in selecting who has access to a cloud environment. Community clouds and hybrid clouds allow for a mixed degree of control and knowledge of other consumers.

OMB Has Undertaken Initiatives and Issued Guidance to Increase Agency Adoption of Cloud Computing Services

According to OMB, the federal government needs to shift from building custom computer systems to adopting cloud technologies and shared services, which will improve the government's operational efficiencies and result in substantial cost savings. To help agencies achieve these benefits, OMB required agencies in 2010 to immediately shift to a "Cloud First" policy and increase their use of available cloud and shared services whenever a secure, reliable, and cost-effective cloud service exists.[7]

---

[7] OMB, *25 Point Implementation Plan to Reform Federal Information Technology Management* (Washington, D.C.: Dec. 9, 2010).

In February 2011, OMB issued the *Federal Cloud Computing Strategy*,[8] as called for in its *25-Point Plan*. The strategy provided definitions of cloud computing services; benefits of cloud services, such as accelerating data center consolidations; a decision framework for migrating services to a cloud environment;[9] case studies to support agencies' migration to cloud computing services; and roles and responsibilities for federal agencies. For example, the strategy stated that NIST's role is to lead and collaborate with federal, state, and local government agency chief information officers, private sector experts, and international bodies to identify standards and guidance and prioritize the adoption of cloud computing services. In addition, the strategy stated that agency cloud service contracts should include SLAs designed to meet agency requirements.

In a December 2011 memo, OMB established the Federal Risk and Authorization Management Program (FedRAMP),[10] a government-wide program intended to provide a standardized approach to security assessment, authorization,[11] and continuous monitoring for cloud computing products and services.[12] All federal agencies must meet FedRAMP requirements when using cloud services and the cloud service providers must implement the FedRAMP security requirements in their cloud environment. To become authorized, cloud service providers provide a security assessment package to be reviewed by the FedRAMP

---

[8]OMB, *Federal Cloud Computing Strategy* (Washington, D.C.: Feb. 8, 2011).

[9]The decision framework, among other things, identifies several key areas for determining the readiness for moving to a cloud environment, including the ability of the cloud service provider to address government security requirements.

[10]OMB, *Security Authorization of Information Systems in Cloud Computing Environments* (Washington, D.C.: Dec. 8, 2011).

[11]Security authorization is the official management decision given by a senior official of an organization to authorize operation of an information system and to explicitly accept the risk to organizational operations and assets, individuals, other organizations, and the nation, based on the implementation of an agreed-on set of security controls.

[12]FedRAMP's security assessment framework encompasses four process areas (document, assess, authorize, and monitor) that are based on the six steps within the framework described in NIST's *Guide for Applying the Risk Management Framework to Federal Information Systems: A Security Life Cycle Approach*, SP 800-37, Revision 1 (Gaithersburg, MD.: February 2010).

Joint Authorization Board,[13] which may grant a provisional authorization. Federal agencies can leverage cloud service provider authorization packages for review when granting an agency authority to operate, where this reuse is intended to save time and money.

Further, at the direction of OMB, the Chief Information Officers Council and the Chief Acquisition Officers Council issued, in February 2012, guidance to help agencies acquire cloud services.[14] In particular, the guidance highlights that SLAs are a key factor for ensuring the success of cloud based services and that federal agencies should include an SLA when creating a cloud computing contract or as a reference. The guidance provides important areas of an SLA to be addressed; for example, it states that an SLA should define performance with clear terms and definitions, demonstrate how performance is being measured, and identify what enforcement mechanisms are in place to ensure the conditions are being met.

In addition, NIST, in its role designated by OMB in the Federal Cloud Computing Strategy, collaborated with private sector organizations to release cloud computing guidance,[15] which affirms the importance of using an SLA when acquiring cloud computing services.

Moreover, a number of other public and private sector organizations have issued research on the incorporation of an SLA in a cloud computing

[13]The Joint Authorization Board is composed of the chief information officers from DOD, DHS, and the General Services Administration and establishes the baseline controls for FedRAMP and criteria for accrediting third-party independent assessment organizations.

[14]Chief Information Officer's Council and Chief Acquisition Officers Council, *Creating Effective Cloud Computing Contracts for the Federal Government*, *Best Practices for Acquiring IT as a Service*.

[15]See, for example, NIST, *US Government Cloud Computing Technology Roadmap Volume 1: High-Priority Requirements to Further USG Agency Cloud Computing Adoption*, NIST SP 500-293, (Gaithersburg, MD.: October 2014); and NIST, *Cloud Computing Synopsis and Recommendations*, NIST SP 800-146, (Gaithersburg, MD.: May 2012).

contract.[16] According to these studies, an SLA is important because it ensures that services are being performed at the levels specified in the cloud computing contract, can significantly contribute to avoiding conflict, and can facilitate the resolution of an issue before it escalates into a dispute. The studies also highlight that a typical SLA describes levels of service using various attributes such as availability, serviceability or performance, and specifies thresholds and financial penalties associated with a failure to comply with these thresholds.

Agencies Are Taking Steps to Implement Prior GAO-Identified Improvements for Cloud-based Computing Services

We have previously reported on federal agencies' efforts to implement cloud computing services and on progress oversight that agencies have made to help federal agencies in those efforts. These include

- In May 2010, we reported on the efforts of multiple agencies to ensure the security of government-wide cloud computing services.[17] We noted that, while OMB, the General Services Administration (GSA), and NIST had initiated efforts to ensure secure cloud computing services, OMB had not yet finished a cloud computing strategy; GSA had begun a procurement for expanding cloud computing services for its website that served as a central location for federal agencies to purchase cloud services, but had not yet developed specific plans for establishing a shared information security assessment and authorization process; and NIST had not yet issued cloud-specific security guidance. We recommended that OMB establish milestones to complete a strategy for federal cloud computing and ensure it addressed information security challenges. These include having a process to assess vendor compliance with government information security requirements and division of information security responsibilities between the customer and vendor.

---

[16]See, for example, European Commission Directorate General for Communications Networks, Content and Technology, *Cloud Service Level Agreement Standardization Guidelines* (Brussels, Belgium: June 2014); MITRE, *Cloud SLA Considerations for the Government Consumer* (September 2012); Cloud Standards Customer Council, *Practical Guide to Cloud Service Agreements*, Version 2.0 (April 2015); and Gartner, *Best Practices for Service Level Agreements for Software as a Service* (Stamford, CT: November 2010).

[17]GAO, *Information Security: Federal Guidance Needed to Address Control Issues with Implementing Cloud Computing*, GAO-10-513 (Washington, D.C.: May 27, 2010).

OMB agreed with our recommendations and subsequently published a strategy in February 2011 that addressed the importance of information security when using cloud computing, but it did not fully address several key challenges confronting agencies, such as the appropriate use of attestation standards for control assessments of cloud computing service providers, and division of information security-related responsibilities between customer and provider. We also recommended that GSA consider security in its procurement for cloud services, including consideration of a shared assessment and authorization process. GSA generally agreed with our recommendations and has since developed the FedRAMP program. Finally, we recommended that NIST issue guidance specific to cloud computing security. NIST agreed with our recommendations and has since issued multiple publications that address such guidance.

- In April 2012, we reported that more needed to be done to implement OMB's *25-Point Plan* and measure its results.[18] Among other things, we reported that, of the 10 key action items that we reviewed, 3 had been completed and 7 had been partially completed by December 2011. In particular, OMB and agencies' cloud-related efforts only partially addressed requirements. Specifically, agencies' plans were missing key practices, such as a discussion of needed resources, a migration schedule, and plans for retiring legacy systems. As a result, we recommended, among other things, that the Secretaries of Homeland Security and Veterans Affairs, and the Attorney General direct their respective CIOs to complete practices missing from the agencies' plans for migrating services to a cloud computing environment. Officials from each of the agencies generally agreed with our recommendations and have taken steps to implement them.

- In July 2012, we reported on the efforts of seven agencies to implement three services by June 2012, including the challenges associated with doing so.[19] Specifically, we reported that selected federal agencies had made progress in implementing OMB's "Cloud First" policy. Seven agencies had implemented 21 cloud computing solutions and had spent a total of $307 million for cloud computing in

---

[18]GAO, *Information Technology Reform: Progress Made; More Needs to Be Done to Complete Actions and Measure Results*, GAO-12-461 (Washington, D.C.: Apr. 26, 2012).

[19]GAO, *Information Technology Reform: Progress Made but Future Cloud Computing Efforts Should Be Better Planned*, GAO-12-756 (Washington, D.C.: July 11, 2012).

fiscal year 2012, about 1 percent of their total IT budgets. While each of the seven agencies had submitted plans to OMB for implementing their cloud services, a majority of the plans were missing required elements. Agencies also identified opportunities for future cloud service implementations, such as moving storage and help desk services to a cloud environment. Agencies also shared seven common challenges that they experienced in moving services to cloud computing. We made recommendations to the agencies to develop planning information, such as estimated costs and legacy IT systems' retirement plans, for existing and planned services. The agencies generally agreed with our recommendations and have taken actions to implement them.

- In September 2014, we reported on the aforementioned seven agencies' efforts to implement additional cloud computing services, any reported cost savings as a result of implementing those cloud services, and challenges associated with the implementation.[20] All of the seven federal agencies we reviewed had added more cloud computing services; the number of cloud services implemented by them had increased from 21 to 101 between fiscal years 2012 and 2014. In addition, agencies had collectively doubled the percentage of their IT budgets from 1 to 2 percent during the fiscal year 2012–14 period. Further, the agencies reported a collective cost savings of about $96 million through fiscal year 2013. We made recommendations to the agencies to assess their IT investments that had yet to be evaluated for suitability for cloud computing services. For the most part, the agencies generally agreed with our recommendations and have taken actions to implement them.

---

[20]GAO, *Cloud Computing: Additional Opportunities and Savings Need to Be Pursued,* GAO-14-753 (Washington, D.C.: Sept. 25, 2014).

Key Practices for Cloud Computing Service Level Agreements Can Help Agencies Manage Services More Effectively

Based on our analysis of practices recommended by the ten organizations with expertise in the area of SLAs and OMB, we compiled the following list of ten practices that are key for federal agencies to incorporate into a contract to help ensure services are performed effectively, efficiently, and securely for cloud computing services. The key practices are organized by the following management areas—roles and responsibilities, performance measures, security, and consequences. [21]

Table 1: Key Practices for a Cloud Computing Service Level Agreement

Roles and responsibilities

1. Specify roles and responsibilities of all parties with respect to the SLA, and, at a minimum, include agency and cloud providers.

2. Define key terms, such as dates and performance.

Performance measures

3. Define clear measures for performance by the contractor. Include which party is responsible for measuring performance. Examples of such measures would include
   - Level of service (e.g., service availability—duration the service is to be available to the agency).
   - Capacity and capability of cloud service (e.g., maximum number of users that can access the cloud at one time and ability of provider to expand services to more users).
   - Response time (e.g., how quickly cloud service provider systems process a transaction entered by the customer, response time for responding to service outages).

4. Specify how and when the agency has access to its own data and networks. This includes how data and networks are to be managed and maintained throughout the duration of the SLA and transitioned back to the agency in case of exit/termination of service.

5. Specify the following service management requirements:
   - How the cloud service provider will monitor performance and report results to the agency.
   - When and how the agency, via an audit, is to confirm performance of the cloud service provider.

6. Provide for disaster recovery and continuity of operations planning and testing, including how and when the cloud service provider is to report such failures and outages to the agency. In addition, how the provider will remediate such situations and mitigate the risks of such problems from recurring.

---

[21]This report identifies key practices for cloud computing SLAs, including a key practice that specifies consequences for missing SLA performance measures. This key practice is defined only in terms of consequences and available remedies that are specifically linked to the SLA performance measures, and does not refer to enforcement mechanisms (such as termination for default) that are made available in the standard terms of federal government contracts generally.

7.	Describe any applicable exception criteria when the cloud provider's performance measures do not apply (e.g., during scheduled maintenance or updates).
Security	
8.	Specify metrics the cloud provider must meet in order to show it is meeting the agency's security performance requirements for protecting data (e.g., clearly define who has access to the data and the protections in place to protect the agency's data).
9.	Specifies performance requirements and attributes defining how and when the cloud service provider is to notify the agency when security requirements are not being met (e.g., when there is a data breach).
Consequences	
10.	Specify a range of enforceable consequences, such as penalties, for non-compliance with SLA performance measures.

Source: GAO analysis of data from public and private organizations. I GAO-16-325

Roles and responsibilities: (1) Define the roles and responsibilities of the major stakeholders involved in the performance of the SLA and cloud contract. These definitions would include, for example, the persons responsible for oversight of the contract, audit, performance management, maintenance, and security. (2) Define key terms, including activation date, performance, and identify any ambiguities in the definitions of cloud computing terms in order to provide the agency with the level of service they can expect from their cloud provider. Without clearly defined roles, responsibilities, and terms, the agency may not be able to appropriately measure the cloud provider's performance.

Performance measures: (1) Define the performance measures of the cloud service, including who is responsible for measuring performance. These measures would include, among other things, the availability of the cloud service; the number of users that can access the cloud at any given time; and the response time for processing a customer transaction. Providing performance parameters provides both the agency and service provider with a well-defined set of instructions to be followed. (2) Specify how and when the agency would have access to its data, including how data and networks will be managed and maintained throughout the life cycle of the service. Provide any data limitations, such as who may or may not have access to the data and if there are any geographic limitations. (3) Specify management requirements, for example, how the cloud service provider would monitor the performance of the cloud, report incidents, and how and when they would plan to resolve them. In addition, identify how and when the agency would conduct an audit to monitor the performance of the service provider, including access to the provider's performance logs and reports. (4) Provide for disaster recovery and continuity of operations planning and testing. This includes, among other things, performing a risk management assessment; how the cloud service would be managed by the provider in the case of a disaster; how data would be recovered; and what remedies would apply during a service failure. (5) Describe applicable exception criteria for when the cloud provider's service performance measures do not apply, such as during scheduled cloud maintenance or when updates occur. Without any type of performance measures in place, agencies would not be able to determine whether the cloud services under contract are meeting expectations.

Security: (1) Specify the security performance requirements that the service provider is to meet. This would include describing security performance metrics for protecting data, such as data reliability, data preservation, and data privacy. Cleary define the access rights of the cloud service provider and the agency as well as their respective responsibilities for securing the data, applications, and processes to meet all federal requirements. (2) Describe what would constitute a breach of security and how and when the service provider is to notify the agency when the requirements are not being met. Without these safeguards, computer systems and networks as well as the critical operations and key infrastructures they support may be lost, and information—including sensitive personal information—may be compromised, and the agency's operations could be disrupted.

Consequences: Specify a range of enforceable consequences, including the terms under which a range of penalties and remedies would apply for non-compliance with the SLA performance measures. Identify how such enforcement mechanisms would be imposed or exercised by the agency. Without penalties and remedies, the agency may lack leverage to enforce compliance with contract terms when situations arise.

OMB Guidance Addresses Seven of the Ten Key Practices

Guidance issued in February 2012, at the direction of OMB highlighted SLAs as being a key factor for ensuring the success of cloud-based services and advised that federal agencies should include an SLA or a reference within the contract when creating a cloud computing contract. The guidance provides areas of an SLA to be addressed; for example, it states that an SLA should define performance with clear terms and definitions, demonstrate how performance is being measured, and identify what enforcement mechanisms are in place to ensure the conditions are being met.

However, the guidance addressed only seven of the ten key practices listed in table 1 that could help agencies better track performance and thus ensure the effectiveness of their cloud services. Specifically, the guidance did not specify how and when the agency would have access to its data, provide for disaster recovery and continuity of operations planning, and describe any exception criteria.

OMB staff members said that, although the guidance drafted by the Chief Information Officers Council and the Chief Acquisition Officers Council was a good start, including all ten key practices should be considered. Without complete guidance from OMB, there is limited assurance that agencies will apply all the key SLA practices into their cloud computing contracts, and therefore may be unable to hold contractors accountable when performance falls short of their goals.

Selected Agencies Incorporated Most of the Key Practices, but Differed in Addressing Them

Many of the 21 cloud service contracts we reviewed at the five selected agencies incorporated a majority of the key practices, but the number of practices differed among contracts. Specifically, seven of the cloud service contracts reviewed met all 10 of the key practices. This included three from DHS, three from Treasury, and one from VA. The following figure shows the total cloud service contracts reviewed and the number that met the 10 key practices at the five selected agencies.

Figure 2: Number of Cloud Service Contracts That Met All 10 Key Practices at Selected Agencies

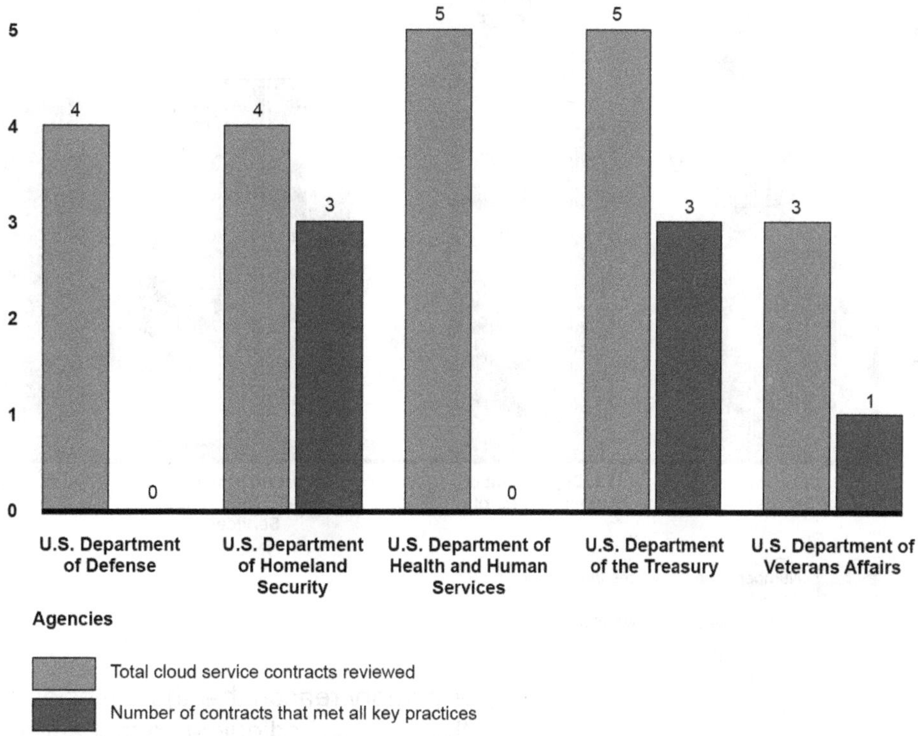

Total cloud service contracts reviewed

Number of contracts that met all key practices

Source: GAO analysis of agency data. | GAO-16-325

Of the remaining 14 cloud service contracts, 13 incorporated five or more of the key practices, and 1 did not meet any of the key practices. Figure 3 shows each of the cloud service contracts we reviewed and the extent to which the agency had included key practices in its SLA contracts. Appendix II includes our analysis of all the cloud services we reviewed, by agency.

Figure 3: Number of 10 Key Practices Met on 21 Cloud Service Contracts at Selected Agencies

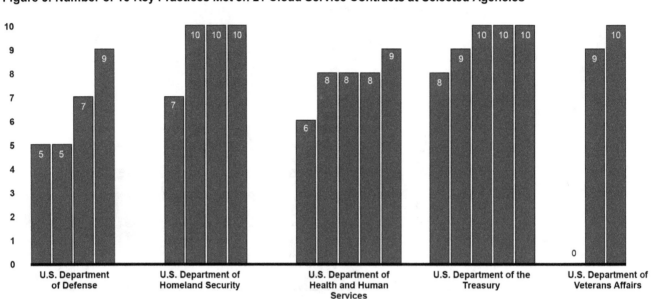

Number of key practices that were met

Source: GAO analysis of agency data. | GAO-16-325

A primary reason that the agencies did not include all of the practices was that they lacked guidance that addresses these SLA practices. Of the five agencies, only DOD had developed cloud service contracting guidance that addressed some of the practices. More specifically, DOD's guidance only addressed three of the key practices: disaster recovery and continuity of operations planning, metrics on security performance requirements, and notifying the agency when there is a security breach. In addition, the guidance partially addressed the practice on access to agency data, specifically, with regard to transitioning data back to the agency in case of exit/termination of service.

Agency officials responsible for the cloud services that did not meet or only partially met key practices provided the following additional reasons for not including all ten practices:

- Officials from DOD's Office of the Chief Information Officer told us that the reason key practices were not always fully addressed is that, when the contracts and associated SLAs were developed, they did not have the aforementioned DOD guidance on cloud service acquisition and use—namely, the agency's memorandum on acquiring cloud services that was released in December 2014,[22] and the current *Defense Federal Acquisition Regulation Supplement*, which was finalized in August 2015.[23] However, as previously stated, this updated guidance addressed three of the ten key practices, and part of one other.

- Officials from DHS's Office of the Chief Information Officer stated that the Infrastructure as a Service cloud service addressed the partially met and not met key practices but did not provide supporting documentation to show that the practices were in place. If key practices have not been incorporated, the system may have decreased performance and the cloud service may not meet its intended goals.

- HHS officials from the National Institutes of Health attributed unmet or partially met practices for four cloud services—Remedy Force, Medidata, the BioMedical Imaging and BioEngineering website, and the Drug Abuse public website—to the fact that they evaluate the cloud vendor's ability to meet defined agency needs, rather than negotiate with vendors on SLA requirements. While this may explain their shortfalls in not addressing all SLA key practices, the agency may be placing their systems at risk of not conducting adequate service level measurements, which may result in decreased service levels.

  HHS officials from the Administration of Children and Families stated that the reason key practices were partially addressed or not addressed for the Grant Solutions cloud service was that these practices were being managed by HHS personnel using other tools and plans, rather than via the SLA established for this service. For example, according to the officials, they are using a management information system to monitor performance of the cloud provider. In addition, with respect to disaster management, the officials said that they have their own disaster recovery plan. Nonetheless, leading studies show that these practices should still be incorporated as part of the cloud service contract to ensure agencies have the proper control over their cloud services.

---

[22]DOD Memorandum, *Updated Guidance on the Acquisition and Use of Commercial Cloud Computing Services* (December 15, 2014).

[23]80 Fed. Reg. 51739 (August 26, 2015), adding 48 C.F.R. Subpart 239.76.

- Treasury officials said the reason, among other things, the SLAs for Treasury Web Services and IRS Portal Environment only partially met certain key practices was because the practices were being provided by support contractors hired by the cloud service provider, and were not directly subject to the SLAs established between Treasury and the cloud service provider. Nonetheless, while having contractors perform practices is an acceptable approach, Treasury officials were unable to provide supporting documentation to show that support contractors were assisting with the practices in question.

- Officials from VA's Office of Information and Technology said the reason the key practice associated with penalties and remedies was not included in the Terremark SLA was because penalties were addressed within other parts of the contract; however, officials were not able to provide documentation identifying such penalties. With regard to an SLA for eKidney, officials told us they had not addressed any of the key practices due to the fact that an SLA was not developed between the agency and cloud service provider. Without including an SLA in cloud service contracts, the agency runs the risk of not having the mechanisms in place to effectively evaluate or control contractor performance.

Until these agencies develop SLA guidance and incorporate all key practices into their cloud computing contracts, they may be limited in their ability to measure the performance of the services, and, therefore, may not receive the services they require.

Conclusions

Although OMB has provided agencies guidance to better manage contracts for cloud computing services, this guidance does not include all the key practices that we identified as necessary for effective SLAs. Similarly, Defense, Homeland Security, Health and Human Services, Treasury, and Veterans Affairs have incorporated many of the key practices in the cloud service contracts they have entered into. Overall, this is a good start towards ensuring that agencies have mechanisms in place to manage the contracts governing their cloud services. However, given the importance of SLAs to the management of these million-dollar service contracts, agencies can better protect their interests by incorporating the pertinent key practices into their contracts in order to ensure the delivery and effective implementation of services they contract for. In addition, agencies can improve management and control over their cloud service providers by implementing all recommended and applicable SLA key practices.

Recommendations for Executive Action

- To ensure that agencies are provided with more complete guidance for contracts for cloud computing services, we recommend that the Director of OMB include all ten key practices in future guidance to agencies.

- To help ensure continued progress in the implementation of effective cloud computing SLAs, we recommend that the Secretary of Defense direct the appropriate officials to ensure key practices are fully incorporated for cloud services as the contracts and associated SLAs expire. These efforts should include updating the DOD memorandum on acquiring cloud services and current *Defense Acquisition Regulations System* to more completely include the key practices.

- To help ensure continued progress in the implementation of effective cloud computing SLAs, we recommend that the Secretaries of Health and Human Services, Homeland Security, Treasury, and Veterans Affairs direct appropriate officials to develop SLA guidance and ensure key practices are fully incorporated as the contract and associated SLAs expire.

Agency Comments and Our Evaluation

In commenting on a draft of this report, four of the agencies—DOD, DHS, HHS, and VA—agreed with our recommendations; and OMB and one agency (Treasury) had no comments. The specific comments from each agency are as follows:

- In an e-mail received on March 25, 2016, OMB staff from the Office of E-Government and Information Technology stated that the agency had no comments at this time.

- In written comments, the Department of Defense concurred with our recommendation and described actions it plans to take to address the recommendation. Specifically, DOD stated that it will update its cloud computing guidance and contracting guidance as appropriate. The Department of Defense's comments are reprinted in appendix III.

- In written comments, the Department of Homeland Security concurred with our recommendation and described actions it plans to take to address the recommendation. Specifically, the department will establish common cloud computing service level agreement guidance. DHS also provided technical comments, which we have incorporated in the report as appropriate. The Department of Homeland Security's comments are provided in appendix IV.

- In written comments, the Department of Health and Human Services concurred with our recommendation, but noted that it was not directed by a federal mandate. We acknowledge that our recommendation is not directed by a mandate; however, implementing leading practices for cloud computing can result in significant benefits. The department also provided technical comments, which we have incorporated in the report as appropriate. The Department of Health and Human Service's comments are provided in appendix V.

- In an e-mail received on March 18, 2016, an audit liaison from the Department of the Treasury's Office of the CIO stated that the department had no comment.

- In written comments, the Department of Veterans Affairs concurred with our recommendation and described planned actions to address it. For example, the department will develop service level agreement guidance to include the 10 key practices. The Department of Veterans Affairs comments are provided in appendix VI.

We are sending copies of this report to interested congressional committees; the Secretaries of Defense, Health and Human Services, Homeland Security, the Treasury, and Veterans Affairs; and the Director of the Office of Management and Budget, and other interested parties. This report will also be available at no charge on our website at http://www.gao.gov.

If you or your staffs have any questions on matters discussed in this report, please contact me at (202) 512-9286 or pownerd@gao.gov. Contact points for our Offices of Congressional Relations and Public Affairs may be found on the last page of this report. GAO staff who made major contributions to this report are listed in appendix VII.

David A. Powner
Director
Information Technology Management Issues

Appendix I: Objectives, Scope, and Methodology

Our objectives were to (1) identify key practices used in cloud computing service level agreements (SLA) to ensure service is performed at specified levels and (2) determine the extent to which federal agencies have incorporated such practices into their cloud computing service level agreements.

To identify key practices used in cloud computing service level agreements, we analyzed SLA research, studies, and guidance developed and used by federal agencies and private entities.
We then performed a comparative analysis of the practices to identify the practices that were recommended by at least two sources. Specifically, we analyzed information from publications and related documentation issued by the following ten public and private organizations to determine key SLA practices:

- Federal Chief Information Officer Council
- Chief Acquisitions Officers Council
- National Institute of Standards and Technology
- European Commission Directorate General for Communications Networks, Content and Technology
- Office of Management and Budget
- Gartner
- MITRE Corporation
- Cloud Standards Customer Council
- International Organization for Standardization
- International Electrotechnical Commission

Next, we organized these practices into management areas and validated our analysis through interviews with experts from these organizations. We also had officials from the Office of Management and Budget (OMB) review and validate that these practices are the ones the office expects federal agencies to follow. In cases where experts disagreed, we analyzed their responses, including the reasons they disagreed, and made changes as appropriate. These actions resulted in our list of key practices for cloud service SLAs.

To determine the extent to which federal agencies have incorporated key practices into their cloud computing contracts, we selected five agencies to review based, in part, on those with the largest fiscal year 2015 IT budgets and planned spending on cloud computing services. The agencies selected were the Departments of Defense (DOD), Homeland Security (DHS), Health and Human Services (HHS), Treasury, and Veterans Affairs (VA).

We selected these agencies based on the following two factors. First, they have the largest planned IT budgets for fiscal year 2015. Their budgets, which collectively totaled $57 billion, represent about 72 percent of the total federal IT budget ($78 billion). Second, these agencies plan to spend relatively large amounts on cloud computing. Specifically, based on our analysis of OMB's fiscal year 2015 budget data, each of the five departments were in the top 10 for the largest amount budgeted for cloud computing and collectively planned to spend $1.2 billion on cloud computing, which represents about 57 percent of the total amount that federal agencies plan to invest in cloud computing ($2.1 billion).

To select and review the cloud services used by the agencies, we obtained an inventory of cloud services for each of the five agencies, and then, for each agency, we listed their cloud services in a random fashion and selected the first two cloud services in the list for each of the three major cloud service models (infrastructure, platform, and software). In certain cases, the agency did not have two cloud services for a service model, so the number chosen for that service model was less than two. This resulted in a non-generalizable sample of 23 cloud services. However, near the end of our engagement, agencies identified 2 of the services as being in a pilot stage (one from DHS, and one from HHS), and thus not operational. We excluded these services from our analysis, as our methodology to only assess operational cloud services. Due to the stage of the engagement, we were unable to select additional services for review. Further, because no computer-generated data was used we determined that there were no data reliability issues.

For each of the selected services, we compared its cloud service contract (if one existed) and any associated SLA documentation to our list of key practices to determine if there were variances and, if so, their cause and impact. To do so, two team analysts independently reviewed the cloud service contracts against the key practices using the following criteria:

- Met: all aspects of the key practices were fully addressed.
- Partially met: some key practices were addressed.
- Did not meet: no key practices were addressed.

In cases where analysts differed on the assessments, we discussed what the rating should be until we reached a consensus. We also interviewed agency officials to corroborate our analysis and identify the causes and impacts of any variances.

We conducted this performance audit from January 2015 to April 2016 in accordance to generally accepted government auditing standards. Those standards require that we plan and perform the audit to obtain sufficient, appropriate evidence to provide a reasonable basis for our findings and conclusions based on our audit objectives. We believe that the evidence obtained provides a reasonable basis for our findings and conclusions based on our audit objectives.

Appendix II: Analysis of Agencies' Cloud Service SLAs against Key Practices

The following tables show each of the five agencies'—DOD, DHS, HHS, Treasury, and VA—cloud services we assessed and our analysis of each contract for cloud services against the key practices. In cases where the SLA partially met a practice, the analysis also includes discussion of the rationale for why that assessment was provided.

(Note: M – met, P – partially met, NM – did not meet.)

Table 2: Analysis of DOD's Cloud Service SLAs against Key Practices

DOD cloud services	Air Force, Integrated Risk Information System			Air Force, Case Tracking			Army, Email as a Service			Navy, Navy Web Portal		
Key practices	M	P	NM	M	P	NM	M	P	NM	M	P	NM
Stakeholder roles and responsibilities	X			X			X			X		
Terms and dates	X			X			X			X		
Measurable performance objectives	X			X			X			X		
Access to agency data		X		X			X				X	
Service management requirements			X	X				X				X
Disaster recovery planning			X	X					X			X
Exception criteria	X			X			X			X		
Security performance requirements			X		X			X			X	
Notification of security breach			X	X			X					X
Consequences	X			X			X			X		
Total	**5**	**1**	**4**	**9**	**1**		**7**	**2**	**1**	**5**	**2**	**3**

Source: GAO analysis of DOD data. I GAO-16-325

With regard to those services that partially met key practices:

Air Force

- The Integrated Risk Information System partially addressed one key practice on how and when the agency was to have access to its data and networks. It included how the data would be transitioned, but did not specify how access to data and networks was to be managed or maintained.

- The Case Tracking cloud service partially included the practice on specifying metrics for security performance requirements. It specified how security needs were to be met but did not give specific metrics for doing so.

Army

- Email as a Service partially addressed two key practices. For the practice on specifying service management requirements, it specified how the cloud service provider was to monitor performance, but did not address how the provider was to report performance or how the agency was to confirm the performance. For the other practice on specifying metrics for security performance requirements, it included how security needs were to be met but did not specify the security metrics.

Navy

- The Web Portal partially incorporated two key practices. For the practice on how and when the agency was to have access to its data and networks, it specified how the data was to be transitioned, but not how access to data and networks was to be managed or maintained. For the other practice on specifying metrics for security performance requirements, it included monitoring of the contractor regarding security, but did not specify security metrics.

Table 3: Analysis of DHS's Cloud Service SLAs against Key Practices

DHS Cloud Services	Business Intelligence as a Service			E-mail as a Service			Infrastructure as a Service			Web Content Management as a Service		
Key practice	M	P	NM	M	P	NM	M	P	NM	M	P	NM
Stakeholder roles and responsibilities	X			X			X			X		
Terms and dates	X			X			X			X		
Measurable performance objectives	X			X			X			X		
Access to agency data	X			X				X		X		
Service management requirements	X			X				X		X		
Disaster recovery planning	X			X			X			X		
Exception criteria	X			X			X			X		
Security performance requirements	X			X			X			X		
Notification of security breach	X			X					X	X		
Consequences	X			X			X			X		
Total	10			10			7	2	1	10		

Source: GAO analysis of DHS data. I GAO-16-325

Infrastructure as a Service partially incorporated two key practices. For the practice on how and when the agency was to have access to its data and networks, it specified how and when the agency was to have access to its data and networks, but did not provide how data and networks was to be transitioned back to the agency in case of an exit. For the other practice on service management requirements, it described how the cloud service is to monitor performance, but did not specify how and when the agency was to confirm audits of the service provider's performance.

Table 4: Analysis of HHS's Cloud Service SLAs against Key Practices

HHS Cloud Service	Remedy Force			Medidata Rave			National Institute of Bio-Medical Imaging and Bio-Engineering Public Website			National Institute on Drug Abuse Pubic Website			Grant Solutions		
Key practice	M	P	NM	M	P	NM	M	P	NM	M	P	NM	M	P	NM
Stakeholder roles and responsibilities	X			X			X			X			X		
Terms and dates	X			X			X			X			X		
Measurable performance objectives		X			X		X			X			X		
Access to agency data	X			X			X				X		X		
Service management requirements			X	X			X			X				X	
Disaster recovery planning	X			X			X			X					X
Exception criteria	X			X			X			X			X		
Security performance requirements	X			X			X			X					X
Notification of security breach	X			X			X			X					X
Consequences	X				X				X		X		X		
Total	8	1	1	8	2		9		1	8	2		6	1	3

Source: GAO analysis of HHS data. I GAO-16-325

With regard to those services that partially met key practices,

- National Institute of Health's Remedy Force partially addressed one key practice on defining measurable performance objectives. It included various performance objectives, such as levels of service and availability of the cloud service, capacity and capability, and measures for response time, but it did not include which party was to be responsible for measuring performance.

- The National Institute of Health's Medidata Rave partially incorporated two key practices. It defined measurable performance objectives, specifically it specified levels of service, capacity and capability of the service, and response time, but did not specify the period of time that it was to be measured. For the other practice on specifying a range of enforceable consequences, it specified remedies, but did not identify any penalties related to non-compliance with performance measures.

- The National Institute on Drug Abuse Public Website partially addressed two key practices. For the practice on specifying how and when the agency is to have access to its data and networks, it specified how and when the agency was to have access to its data and networks, but did not identify how data and networks were to be managed throughout duration of the SLA. For the other practice on specifying a range of enforceable consequences, it included a number of remedies, but did not specify a range of enforceable penalties.

- HHS's Grant Solutions partially incorporated one key practice on specifying service management requirements. It provided for when and how the agency was to confirm cloud provider performance, but did not specify how the cloud service provider was to monitor performance and report results.

Table 5: Analysis of Treasury's Cloud Service SLAs against Key Practices

Treasury Cloud Services	Enterprise Asset Management System			Franchise Financial and Administrative Services			IRS Portal Environment			Manufacturing Support Suite			Treasury Web Solutions		
Key Practices	M	P	NM	M	P	NM	M	P	NM	M	P	NM	M	P	NM
Stakeholder roles and responsibilities	X			X			X			X			X		
Terms and dates	X			X			X			X			X		
Measurable performance objectives	X			X			X			X			X		
Access to agency data	X			X				X		X					X
Service management requirements	X			X			X			X			X		
Disaster recovery planning	X			X			X			X			X		
Exception criteria	X			X			X			X			X		
Security performance requirements	X			X			X			X			X		
Notification of security breach	X			X			X			X			X		
Consequences	X			X			X			X				X	
Total	**10**			**10**			**9**	**1**		**10**			**8**	**2**	

With regard to those services that partially met key practices,

- Treasury's Internal Revenue Service's Portal Environment partially included one key practice on specifying how and when the agency was to have access to its data and networks. It specified how and when the agency was to have access to its data and networks, but it did not provide on how data and networks were to be transitioned back to the agency in case of an exit.

- The Treasury's Web Solutions partially addressed two key practices. For the practice on specifying how and when the agency was to have access to its data and networks, it specified how and when the agency was to have access to its data and networks, but it did not provide how data and networks would be transitioned back to the agency in case of an exit. For the other practice on specifying a range of enforceable consequences, it did not provide detailed information on a range of enforceable penalties and remedies for non-compliance with SLA performance measures.

Table 6: Analysis of VA's Cloud Service SLAs against Key Practices

VA Cloud Services	Customer Relationship Management			eKidney			Terremark		
Key Practices	M	P	NM	M	P	NM	M	P	NM
Stakeholder roles and responsibilities	X					X	X		
Terms and dates	X					X	X		
Measurable performance objectives	X					X	X		
Access to agency data	X					X	X		
Service management requirements	X					X	X		
Disaster recovery planning	X					X	X		
Exception criteria	X					X	X		
Security performance requirements	X					X	X		
Notification of security breach	X					X	X		
Consequences	X					X			X
Total	10					10	9		1

Source: GAO analysis of VA data. I GAO-16-325

Appendix III: Comments from the Department of Defense

DEPARTMENT OF DEFENSE
6000 DEFENSE PENTAGON
WASHINGTON, D.C. 20301-6000

CHIEF INFORMATION OFFICER

David A. Powner
Director of IT Management Issues
U.S. Government Accountability Office
441 G. Street, NW
Washington, DC 20548

MAR 2% 2016

Dear Mr. Powner:

Enclosed please find the Department of Defense (DoD) response to the GAO Draft Report, GAO-16-325, "CLOUD COMPUTING: Agencies Need to Incorporate Key Practices to Ensure Effective Performance" dated February 19, 2016 (GAO Code 311411).

Thank you for the opportunity to comment on this draft report. DoD appreciates the work performed by the Government Accountability Office (GAO) defining key practices in cloud computing service level agreements (SLAs) and determining the extent to which federal agencies have incorporated such practices into their SLAs. Adoption of cloud computing is rapidly expanding within the Federal Government and efforts such as this one are critical to ensure the performance, dependability and security of the Federal information technology (IT) environment.

DoD is pleased to note GAO's recognition of the work already done in the Department to address key practices in contracting guidance and cloud computing policies. This guidance ensures that the Department's cloud contracts address the critical practices of disaster recovery and continuity of operations planning, metrics on security performance requirements, and notifying the agency when there is a security breach. DoD is committed to working with GAO to ensure the effective performance of cloud computing services.

Should you have any questions, please contact Mr. Robert Vietmeyer, robert.w.vietmeyer.civ@mail.mil, 571-372-4461.

Sincerely,

David L. De Vries
Principal Deputy
Department of Defense Chief Information Officer

Enclosure:
Department Of Defense Comments To The GAO Recommendation

**GAO DRAFT REPORT DATED FEBRUARY 19, 2016
GAO-16-325 (GAO CODE 311411)**

**"CLOUD COMPUTING: AGENCIES NEED TO INCORPORATE KEY PRACTICES
TO ENSURE EFFECTIVE PERFORMANCE"**

**DEPARTMENT OF DEFENSE COMMENTS
TO THE GAO RECOMMENDATION**

RECOMMENDATION: To help ensure continued progress in the implementation of effective cloud computing SLAs, GAO recommends that the Secretary of Defense direct the appropriate officials to ensure key practices are fully incorporated for cloud services as the contracts and associated SLAs expire. These efforts should include updating the DoD memorandum on acquiring cloud services and current Defense Federal Acquisition Regulations, to more completely include the key practices.

DoD RESPONSE: Concur. The Department agrees that the key practices identified by GAO are important considerations and should be included in SLAs. In recognition of this, the DoD CIO will update DoD cloud computing guidance by the end of the calendar year, and work with Defense Acquisition Procurement Policy (DPAP) to update contracting guidance as appropriate. Since the key practices identified by GAO are applicable across the Federal Government, the Federal Acquisition Regulation (FAR) rather than the Defense Acquisition Regulations System (DFARS) may be the best place to address the requirement.

Appendix IV: Comments from the Department of Homeland Security

U.S. Department of Homeland Security
Washington, DC 20528

Homeland Security

March 21, 2016

David A. Powner
Director, Information Technology Management Issues
U.S. Government Accountability Office
441 G Street, NW
Washington, DC 20548

Re: Draft Report GAO-16-325, "CLOUD COMPUTING: Agencies Need to
       Incorporate Key Practices to Ensure Effective Performance"

Dear Mr. Powner:

Thank you for the opportunity to review and comment on this draft report. The U.S.
Department of Homeland Security (DHS) appreciates the U.S. Government
Accountability Office's (GAO's) work in planning and conducting its review and issuing
this report.

The Department is pleased to note GAO's positive recognition that DHS has incorporated
key practices for cloud computing service level agreements (SLAs) within its existing
cloud contracts. DHS is committed to incorporating the use of key practices whenever
possible in its SLAs to help ensure the delivery and effective implementation of the
cloud-based services it contracts for.

The draft report contained one recommendation for DHS with which the Department
concurs. Specifically, GAO recommended that:

Recommendation 1: The Secretary of Homeland Security direct appropriate officials to
develop SLA guidance and ensure key practices are fully incorporated as the contract and
associated SLAs expire.

Response: Concur. DHS will establish common cloud computing SLA guidance to
better ensure the delivery of trusted and quality cloud computing services. The DHS
Office of the Chief Information Officer (OCIO), Enterprise Business Management Office
(EBMO) will lead the activities to address the recommendation. EBMO will facilitate
input from other offices within the OCIO and from DHS Components to establish
standard SLA guidance, circulate the guidance throughout the enterprise, and ensure key
practices are fully incorporated as contracts and associated SLAs expire. Estimated
Completion Date: September 30, 2016.

Again, thank you for the opportunity to review and comment on the draft report. Technical comments were previously provided under separate cover. Please feel free to contact me if you have any questions. We look forward to working with you in the future.

Sincerely,

Jim H. Crumpacker, CIA, CFE
Director
Departmental GAO-OIG Liaison Office

2

Appendix V: Comments from the Department of Health & Human Services

DEPARTMENT OF HEALTH & HUMAN SERVICES   OFFICE OF THE SECRETARY

Assistant Secretary for Legislation
Washington, DC 20201

MAR 2 2 2016

David Powner
Director, Information Technology
U.S. Government Accountability Office
441 G Street NW
Washington, DC 20548

Dear Mr. Powner:

Attached are comments on the U.S. Government Accountability Office's (GAO) report entitled, *"Cloud Computing: Agencies Need to Incorporate Key Practices to Ensure Effective Performance"* (GAO-16-325).

The Department appreciates the opportunity to review this report prior to publication.

Sincerely,

Jim R. Esquea
Assistant Secretary for Legislation

Attachment

<u>**GENERAL COMMENTS OF THE DEPARTMENT OF HEALTH AND HUMAN SERVICES (HHS) ON THE GOVERNMENT ACCOUNTABILITY OFFICE'S (GAO) DRAFT REPORT ENTITLED: CLOUD COMPUTING: AGENCIES NEED TO INCORPORATE KEY PRACTICES TO ENSURE EFFECTIVE PERFORMANCE (GAO-16-325)**</u>

The U.S. Department of Health and Human Services (HHS) appreciates the opportunity from the Government Accountability Office (GAO) to review and comment on this draft report.

<u>**GAO Recommendation**</u>
To help ensure continued progress in the implementation of effective cloud computing service level agreements (SLAs), we recommend that the Secretaries of HHS, the Department of Homeland Security, Treasury, and Veterans Affairs direct appropriate officials to develop SLA guidance and ensure key practices are fully incorporated as the contract and associated SLAs expire.

<u>**HHS Response:**</u>
HHS concurs with GAO's recommendation which is not supported by a federal mandate at the time of this report. HHS also agrees that the leading practices are not always identified in the Service Level Agreements (SLAs), however, we do have other tools to monitor these practices through other mechanisms monitoring that provide performance metrics and analyzes performance in more detail. Once a federal mandate is implemented for all Agencies to comply with the implementation of the stated recommendations, HHS will work with its Operating Divisions to be in compliance.

Appendix VI: Comments from the Department of Veterans Affairs

DEPARTMENT OF VETERANS AFFAIRS
WASHINGTON DC 20420

March 25, 2016

Mr. David Powner
Director
Information Technology Management Issues
U.S. Government Accountability Office
441 G Street, NW
Washington, DC 20548

Dear Mr. Powner:

The Department of Veterans Affairs (VA) has reviewed the Government Accountability Office's (GAO) draft report, *"Cloud Computing: Agencies Need to Incorporate Key Practices to Ensure Effective Performance"* (GAO-16-325). VA generally agrees with GAO's conclusions.

The enclosure sets forth the action to be taken to address the GAO draft report recommendation.

Sincerely,

Robert D. Snyder
Chief of Staff

Enclosure

Enclosure

Department of Veterans Affairs (VA) Response to
Government Accountability Office (GAO) Draft Report
*"CLOUD COMPUTING: Agencies Need to Incorporate Key Practices to Ensure
Effective Performance"*
(GAO-16-325)

GAO Recommendation: **To help ensure continued progress in the implementation
of effective cloud computing SLAs, GAO recommends that the Secretary of
Veterans Affairs direct appropriate officials to develop SLA guidance and ensure
key practices are fully incorporated as the contract and associated SLAs expire.**

VA Comment: Concur. The Department of Veterans Affairs (VA) Office of Information
and Technology (OI&T) will develop service level agreement (SLA) guidance to include
the 10 key practices in cloud computing SLAs identified in the GAO draft report. The
target completion date to develop and issue the SLA guidance VA-wide is July 29,
2016. Once the SLA guidance has been issued, OI&T will ensure the key practices are
fully incorporated as cloud service contracts and associated SLA's expire.

Appendix VII: GAO Contact and Staff Acknowledgments

GAO Contact	David A. Powner, (202) 512-9286 or pownerd@gao.gov
Staff Acknowledgments	In addition to the contact name above, individuals making contributions to this report included Gary Mountjoy (assistant director), Gerard Aflague, Scott Borre, Nancy Glover, Lori Martinez, Tarunkant Mithani, Karl Seifert, and Andrew Stavisky.

www.ingramcontent.com/pod-product-compliance
Lightning Source LLC
Chambersburg PA
CBHW060509060326
40689CB00020B/4685